Mathias K. B. Lüdecke, Martin Budde, Oles Kit, Diana Reckien

Climate Change Scenarios for Hyderabad
Integrating uncertainties and consolidation

Emerging megacities
Dicussion Papers
Edited by Konrad Hagedorn, Christine Werthmann, Dimitrios Zikos, Ramesh Chennamaneni

Humboldt-Universität zu Berlin
Department of Agricultural Economics
Division of Resource Economics
Philippstr. 13, House 12
10115 Berlin

Tel.: +49 (0)30 2093 6305
Fax: +49 (0)30 2093 6497
www.agrar.hu-berlin.de/struktur/institute/wisola/fg/ress
www.sustainable-hyderabad.de

Contact: emerging.megacities@hu-berlin.de

The emerging megacities discussion papers are available at:
www.eh-verlag.de

ISSN print edition 2193-6927

Emerging megacities Discussion Papers are prepared by researchers working on topics in the realm of sustainable development in Megacities of Tomorrow, a research priority by the German Ministry of Education and Research (BMBF). The papers have been peer-reviewed by a board of external reviewers.
Views and opinions expressed do not necessarily represent those of the Division of Resource Economics.
Comments are highly welcome and should be sent directly to the authors.
We welcome contributions on any topics related to Megacities of Tomorrow. Further information on the submission procedure is given at:
www.sustainable-hyderabad.de/emerging-megacities

Lüdecke, Mathias K. B.; Budde, Martin; Kit, Oles; Reckien, Diana

Climate Change Scenarios for Hyderabad
Integrating uncertainties and consolidation

Emerging megacities Discussion Papers, Volume 1/2010

ISBN/EAN: 978-3-86741-818-8

First published in 2012 by Europaeischer Hochschulverlag GmbH & Co KG, Bremen, Germany.

© Europaeischer Hochschulverlag GmbH & Co KG, Fahrenheitstr. 1, D-28359 Bremen (www.eh-verlag.de). All rights reserved.

Cover: Photo "Metropolis", ferendus (flickr). Creative Commons License

No part of this publication may be reproduced or transmitted, in any form or by any means, electronic, mechanical, photocopying, recording or otherwise, or stored in any retrieval system of nay nature, without the written permission of the copyright holder and the publisher, application for which shall be made to the publisher.

EHV

Climate Change Scenarios for Hyderabad

Integrating Uncertainties and Consolidation

Mathias K. B. Lüdecke[*,†] *Martin Budde*[†] *Oles Kit*[†] *Diana Reckien*[†]

September 2010

Abstract

The paper evaluates the global climate projections of 17 AOGCMs with the aim to project 4 climatic variables under two different global emission scenarios (SRES B1 and A2) for the area of Hyderabad (India) for the 21st century. The evaluated model runs were produced within the IPCC AR4 - process. We applied a statistical downscaling to be able to evaluate that large number of model runs. Aggregation of results was done on the basis of model and variable specific weights reflecting the accuracy of reproduction of the current climate. Projection certainty was assessed by the degree of model consensus.

Key words: *climate change, local climate projections, Hyderabad, heat waves, intense rain*

---

[*] Corresponding author. Email: luedeke@pik-potsdam.de
[†] Potsdam Institute for Climate Impact Research (PIK), Research Domain: Climate Impacts and Vulnerabilities, P.O. Box 60 12 03, 14412 Potsdam

1 Introduction

This paper describes the continuation of the work from Lüdeke, Budde et al. (2009). It is about the refinement and consolidation of the projection of climate change signals which the city of Hyderabad and its peri-urban region have to expect.

The climate variables under investigation are:

- The mean annual temperature in the region of Hyderabad, influencing the natural water balance, the peri-urban agriculture and groundwater refill
- The total annual precipitation in the region of Hyderabad, influencing the water supply of the city and the success of peri-urban agriculture
- The frequency distribution of daily precipitation in the urban area, in particular the frequency of heavy rain events
- The frequency and length of heat waves in the city, generating direct health and infrastructure impacts

One objective is to identify the effect of different global emission scenarios on the projected climate variables in Hyderabad. The question, in how far the reduction of global emissions will reduce climate change signals in, e.g., Hyderabad is closely related to efforts one is willing to undertake in contributing to and putting forward these emission reductions. To scan the range of possible future developments of the global greenhouse gas emissions, we choose in this paper the SRES-A2 scenario for a high-emission future and the SRES-B1 scenario for a significant global emission reduction path (SRES, Nakićenović and Swart 2000).

The general methodological idea (for details see Lüdeke, Budde et al., 2009) is to reflect the full range of available global climate change projections (depending on the climate variable and the emission scenario 26-30 different runs generated by 17 climate models) as their differences are decisive for differences of the downscaling results, independent from the chosen method (Hollweg et al. 2008; Orlowsky et al. 2008). So we decided to use relatively simple and fast statistical downscaling algorithms (instead of, e.g., mechanistic, nested high resolution models) to be able to evaluate the full range of Atmosphere-Ocean General Circulation Models (AOGCMs) as a basis for a proper uncertainty analysis.

The main improvements compared to the 2009-report are:

- all available model runs from the AOGCM comparison done for the 4th Assessment Report of the IPCC (Meehl et al. 2007) are considered now

- the different accuracy in reproducing the present climate is used for a weighting of the single AOGCMs in its contribution to the aggregated results
- the downscaling algorithm was adapted to the quality of the available weather data, thereby avoiding artefacts

To our knowledge, this is the first time that the full range of available AOGCM projections is used to assess multivariate local climate change signals by considering a quality measure for the single AOGCMs.

In the following chapters we introduce the applied evaluation algorithms for each of the climate variables, give a short summary of data sources, describe the results, and discuss them with them aim to sketch some policy consequences. In the Appendix all used AOGCM results are documented.

2 Evaluation Algorithms

In the following we first sketch the downscaling and error-correcting algorithms from AOGCM results to the local Hyderabad situation (see also Lüdeke, Budde et al., 2009). Here emphasis is laid on aspects, which had to be changed due to our recent findings of the sensitivity of the results on the quality of the observational data. Furthermore we describe the algorithms for the determination of weights.

The considered variables are the average temperature, the annual precipitation sum, the number of heat wave days per year and the distribution of daily precipitation. For all variables we distinguish between AOGCM model error- and downscaling corrections.

With respect to model errors we assume that the AOGCM results have to represent the averaged climate variables of a larger area (at least in the order of one model grid element). Comparison of model results reproducing the present climate with observational data describing these averages yields (1) a measure for the quality of the model for the considered region and variable and (2) a correction procedure which is applied to the projections of the respective AOGCM. Figure 1 shows the geographical extent of the model grid elements of the considered AOGCMs containing the Hyderabad urban area. For the further analysis the AOGCM results will be compared with climatological values spatially averaged over a rectangular area from 76.25°E, 15.75°N (lower left) –79.75°E, 18.75°N (upper right corner).

With respect to downscaling we follow the approach, which is based on the assumption that the statistical relation between area and point data, which is valid for the present climate can be applied to the projected climate variables as well. Advantages and disadvantages of this statistical downscaling compared to other methods – in scientific and pragmatic terms – is given in Lüdeke, Budde et al. (2009).

Evaluation Algorithm for the Mean Annual Temperature

Here we follow mainly the procedure already used for the previous report. The main difference is that the error correction and the downscaling correction have to be performed as two separate steps to allow for the determination of the weight, which the respective AOGCM is assigned to in the final integration of the results. The areal value (76.25°E, 15.75°N – 79.75°E, 18.75°N) for annual mean temperature (1961-1990) was taken from the CRU_TS 2.1 dataset (Mitchell et al., 2004) and amounted to 27.2°C, being close to the Begumpet station value of 26.9°C. Table 1 gives the variation within this area in the original 0.5°x0.5° resolution of the CRU dataset, ranging from 26.43 to 28.49°C, the Hyderabad area being somewhat closer to the lower bound.

Table 1: Mean annual temperature [°C] for 1961-1990 according to the CRU TS 2.1 dataset. The spatial average is 27.2°C

Latitude	Longitude						
	76.75	*77.25*	*77.75*	*78.25*	*78.75*	*79.25*	*79.75*
18.75	26.96	27.22	27.53	27.24	27.34	27.85	28.05
18.25	26.8	26.96	27.24	26.9	27.01	27.78	27.98
17.75	26.93	26.55	26.47	26.43	26.49	27.56	27.9
17.25	27.39	27.19	26.77	26.59	26.56	27.82	27.98
16.75	27.41	27.37	27.19	26.85	27.12	28.08	28.45
16.25	27.38	27.46	27.73	27.48	26.79	27.47	28.14
15.75	27.38	27.42	27.82	28.00	27.47	27.9	28.49

The AOGCM specific error was calculated as the difference of the mean annual temperature from the model run reproducing the present climate and the observed areal average of 27.2°C. Then the absolute value of this error was taken from each AOGCM and normalised according to the minimum and maximum error: [Error$_{max}$,Error$_{min}$]→[0,1], meaning that the model with the largest error gets the weight 0, the best model the weight 1 while the remaining models lie in between. The resulting weights are summarised in Table 3 for all AOGCMs contributing values for the climate variable under the respective global emission scenario.

To represent the difference between the areal average (which is expected to be reproduced by the AOGCMs) and the value of the variable at Hyderabad, we calculated an

AOGCM-specific scaling factor for the present climate, which maps the AOGCM-value onto the observed value at the Begumpet station. This scaling factor was then fixed and used to downscale the results of the projection runs for the time slices (2046-2065; 2081-2100) and global emission scenarios (B1; A2). In the Appendix we document the original and scaled results of each AOGCM.

Evaluation Algorithm for Annual Precipitation Sum

For the evaluation of the annual precipitation sum we proceed similar as in the above case of annual mean temperature. For the determination of the weight we take the observed areal values (76.25°E, 15.75°N –79.75°E, 18.75°N) for the annual precipitation sum from the same source (see Table 2). The spatially averaged annual precipitation sum amounted to 872 mm, somewhat higher than the respective value reported for the Begumpet station (808 mm).

The AOGCM specific error was calculated and normalised as before, the resulting weights are displayed in the respective column of Table 3. In the Appendix we document the original and scaled precipitation results of each AOGCM.

Table 2: Annual precipitation sums [mm] for 1961-1990 according to the CRU TS 2.1 dataset. The spatial average is 872 mm.

Latitude	Longitude						
	76.75	77.25	77.75	78.25	78.75	79.25	79.75
18.75	899	948	1031	1060	1005	985	1023
18.25	882	949	1035	1069	985	951	985
17.75	843	886	941	944	871	891	955
17.25	814	866	891	840	777	822	916
16.75	752	812	871	829	766	787	879
16.25	673	736	781	763	706	746	824
15.75	602	611	654	682	683	746	826

Evaluation Algorithm for Daily Precipitation

In Lüdeke, Budde et al. (2009) we introduced an advanced method of statistical downscaling for the frequency distribution of daily precipitation, which was based on observed daily time series of point and areal precipitation data. Tests of this algorithm in other world regions showed promising results. At the same time, some doubts about the absolute exactness of the dates of precipitation events, both for the Begumpet weather station record and the areal daily precipitation data set of the Indian Meteorological

Department (IMD) came up. Several shifts of ± one day where detected (via comparison with independent sources like press articles reporting on heavy rain events) and it appeared to be impossible to correct these. Therefore we investigated the sensitivity of our approach to date shifts resulting in the conclusion that the algorithm in Lüdeke, Budde et al. (2009) is not appropriate under the given data uncertainty (in particular for the rather rare events – which are, of course, the interesting ones).

Given this fact we modified our evaluation approach for the change in frequency of days with heavy rain by only applying a distribution dependent and thereby date shift insensitive algorithm, similar to the approach of Tebaldi et al. (2006).

Like in the heat wave case, the weighting of AOGCM quality has to address frequency distribution aspects, here of daily precipitation. For that, we compared the integrated frequency distributions (probability functions) of the areal precipitation time series with the current climate runs of the AOGCMs. We used the Kolmogorov-Smirnov test (Shorak & Wellner, 1986) to quantify the similarity of distribution functions. The applied metricsis the maximum distance of the probability functions. This distance was determined for all AOGCMs and than normalised like in the preceding cases to obtain the weights which are again displayed in the respective column of Table 3.

To evaluate the projection runs of the AOGCMs we first determined the percentiles of interesting precipitation amounts (20 mm/day, 40 mm/day, 80 mm/day, 120 mm/day) in the Begumpet time series of daily precipitation. Then these percentiles were used to determine precipitation limits for each AOGCM's present climate run. These limits were fixed and the change of the frequency in the projection runs was taken as a measure for climate change. The results for each single AOGCM are displayed in the Appendix.

Evaluation Algorithm for Heat Waves

Besides the consideration of all available AOGCMs for the estimation of future heat wave frequencies the main modification compared to Lüdeke, Budde et al. (2009) lies in the heat wave definition which had to be adapted for tropical and subtropical regions.

In its heat wave definition the India Meteorological Department distinguishes between places where the "normal maximum temperature" is more or less than 40°C. Above 40°C "normal maximum temperature" days with maximum temperatures of three to four degrees Celsius above normal are interpreted as "affected by a heat wave" while an exceeding by five degrees Celsius is classified as "severe heat wave". Another category considers regions where the normal maximum temperature is 40°C or less. If in these areas the day temperature is five-six degrees Celsius above normal, then the place is said to be affected by heat waves. A severe heat wave occurs when the day-time temperature exceeds the normal maximum by more than six degrees Celsius (cited after Down to Earth, 2009).

The definition recommended by the World Meteorological Organisation is different and defines a heat wave as the daily maximum temperature of more than five consecutive days which exceed the average maximum temperature by five Celsius degrees, the normal period being 1961–1990 (cited after Frich et al., 2002).

These two examples show that there is no unique definition of heat waves – some emphasising the temperature context, others the duration aspects. Even amongst the Indian authorities there is no unique definition. The "Orissa state disaster management authority", e.g., does not distinguish between the below and above 40°C "normal maximum" temperature but introduces instead an absolute temperature (45°C) which is sufficient to call a day a heat wave day.

We tried to take the basic ideas and key data of these definitions as there is both, diversity and fuzziness in the definitions. In particular the term "normal maximum temperature" is unclear but quantitatively decisive: is it the climatological mean (i.e. the average over 30 years) of: annual/monthly/weakly – maximum/average of daily maximum temperatures? (further interpretations are possible)

The definition of this "normal maximum temperature" is crucial as it characterises the level people are adapted to. Taking this into account, annual averages over maximum daily temperatures are certainly useless even under slight seasonal changes of temperature (which is clearly the case in Hyderabad). So we decided to use a mixture of two elements: the climatological mean of the monthly average of the hottest calendar month (that's the typical daily maximum temperature people expect in that specific hot month) and the climatological mean of daily maximum temperatures of the hottest calendar day in the hottest month. The latter is a measure for the variability people expect in the hottest month. The arithmetic mean of these two temperatures is taken as a measure for the "normal maximum temperature". As threshold for a "heat wave day" we define the daily maximum temperature larger than 3.5°C compared to this normal maximum temperature (this is in the lower range of the above definitions as our definition of normal maximum temperature is rather in the upper range compared to the other possibilities).

Having now clearly defined the "heat wave day" temperature threshold the frequency can be calculated from the observed present day climate time series. On the other hand, the daily T_{max} time series of the model runs for the present climate can also be evaluated along the above heat wave definition and the resulting number of heat wave days per year can be compared to the Hyderabad situation. This difference is a direct measure for the quality of the models with respect to this variable, now emphasising the form of the frequency distribution of the calculated T_{max} values (the absolute value was already assessed in section 2). After normalisation we arrive at the weights, which are again displayed in the respective column of Table 3.

For the projections the heat wave temperature threshold calculated from the present climate run was fixed for each AOGCM and applied to the respective projection runs. As comparable measure we use the absolute change in heat wave days in the future climates. The AOGCM specific results are given in the Appendix.

3 Data Sources and Preparation

In the following section we describe the sources for the observational data (weather station and areal) as well as the model runs. Although data sources did not change much compared to the preceding report, we summarise this part to make the current report readable in a standalone manner.

Observational Data

The first kind of observational data needed are time series of daily precipitation, daily maximum and daily average temperature at a weather station in Hyderabad. The longest time series within the urban area of Hyderabad, at least to our knowledge so far, is the weather station in Begumpet (former international airport), North of the lake Hussein Sagar. Here we could retrieve the almost complete time series for the period 1997 to 2007 from the Xdat – PIK database. Additionally, for the same station we could retrieve the respective data for 1901-1970 from the KNMI-data base, which contains some more gaps but is still very valuable for determining statistical properties.

The second kind of observational data is on areal averages of daily precipitation in the larger area which includes Hyderabad. As the AOGCM results are of this kind, the observational analogue is important for downscaling and calibration. Here the IMD Climate Research Centre proved to be a valuable resource as they could provide areal daily precipitation data for India in a spatial resolution of $1° \times 1°$ for the time period 1951-2007. This data was generated by interpolation of station data and it has to be considered that the quality of this time series varies in time due to changing station density.

Finally we used the data from the IPCC data distribution centre to get the annual mean temperature and the annual precipitation from CRU TS 2.1 (Mitchell et al., 2004) for the area around Hyderabad.

AOGCM Model Runs

The AOGCM-results were taken from the World Climate Research Program's (WCRP's) Coupled Model Inter-comparison Project phase 3 (CMIP3) multi-model dataset which provides, amongst others, the runs performed for the IPCC, AR4 process. We use this archive for the daily data values for precipitation and maximum/average temperature. As already mentioned, we chose the global SRES-emission scenarios A2 and B1 and therefore the runs 20C3M (i.e., the climate of the 20th Century experiment), SRESA2 (i.e., the SRES A2 experiment) and SRESB1 (i.e., the 550 ppm stabilisation experiment). All data was retrieved in netcdf-raster format, including the exact geo-referencing of the data which has a resolution of about $2° \times 2°$.

Data Preparation

Besides the usual homogenisation of the temporal scales of daily time series (representation of leap years etc.) the spatial scales of the areal data sets needed attention (see Figure 1). The observed $1° \times 1°$ areal data was adapted to the AOGCM grid cell by using only these $1° \times 1°$ grid cells which relevantly contribute to the AOGCM grid cell.

4 Results

In the following sections we show the final results for each of the four climate variables: annual average temperature, annual precipitation sum, distribution of daily precipitation – in particular the number of days/year with precipitation amounts greater 80 mm/day – and the number of heat wave days per year. The weighted frequency distributions over the different AOGCMs allow for a detailed interpretation of the consistency of the projections over the different models (and thereby of the certainty of the results) while the tables with the weighted averages of the projected changes and their standard

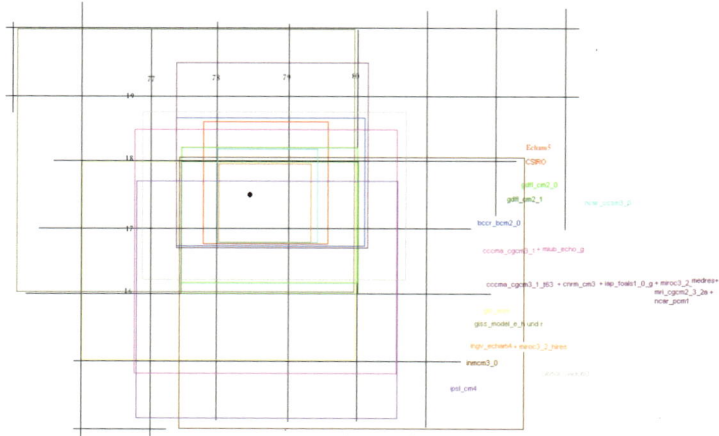

Figure 1: Diagram showing the location of the grid elements best representing the Hyderabad urban area for the considered AOGCMs. The black grid in the background is a 1°x1° reference grid and the black dot shows the location of Hyderabad (Begumpet station).

deviation summarise the analyses. The averages cannot be properly interpreted without taking into consideration the respective frequency distributions.

In Table 3 we display the weighting factors for all AOGCMs used. These weighting factors are limited to the area around Hyderabad and for this reason give no information about the overall performance of the listed AOGCMs. Further it becomes clear that it is not possible to judge on an AOGCM by just one weighting factor because for example the AOGCM cccma_cgcm3_1_t63 reproduces the annual mean temperature and precipitation rather poorly but meets the form of the frequency distribution of daily precipitation and the number of heat wave days quite well.

For a specific climate variable and a specific region the introduced weighting will improve the aggregated evaluation. The procedure to calculate the weighted average is quite straight forward, while the weighted frequency distributions and standard deviations might need some further explanation.

The distribution plots (Figure 2 to Figure 5) are separated in a) the projections under the moderate B1 global emission scenario and b) the projections under the more pessimistic A2 global emission scenario. In each of the plots a black spike over the current climate value gives the reference situation, the small width of the spike symbolises the comparably high certainty of this observational value (although due to measurement uncertainties not totally certain). The two coloured distribution functions reflect the number of models (weighted by their quality) which project a specific climate variable (on the x-axis). So, for a qualitative reading, wide and flat curves hint to a low consensus amongst the different AOGCMs while steep curves reflect relatively high consensus. We follow the notion of the IPCC AR4 that high consensus is an indicator for high certainty of the results. For quantitative reading, the value of the distribution function has to be multiplied with an interval on the x-axis to get the weighted model fraction which projects a value within that interval.

Table 3: Weights attributed to the different AOGCMs for the four climate variables and the two emission scenarios. Blank spaces: respective run is not available.

AOGCM	Scenario	Prc/Year	Days with Prc/Day>80mm	Annual Mean Temperature	Heatwave Days/Year
Bccr_bcm2_0	A2	0.8	1.0	0.4	0.8
	B1	0.8	1.0	0.4	0.8
Cccma_cgcm3_1	A2		1.0	0.2	1.0
	B1		1.0	0.2	1.0
Cccma_cgcm3_1_t63	A2				
	B1	0.3	1.0	0.1	0.9
csiro_mk3_0	A2	0.6	1.0	0.5	0.4
	B1	0.6	1.0	0.5	0.4
csiro_mk3_5	A2	0.8	1.0	0.2	0.1
	B1	0.8	1.0	0.2	0.0
gfdl_cm2_0	A2	0.9	1.0	0.3	0.9
	B1	0.9	1.0	0.3	0.9
gfdl_cm2_1	A2	0.1	0.7	0.0	0.6
	B1	0.2	0.7	0.0	0.5
giss_aom	A2				
	B1	0.9	1.0	0.9	0.8
Giss_model_e_r	A2	0.8	0.0	0.7	0.5
	B1	0.8	0.0	0.7	0.5
iap_fgoals1_0_g	A2				
	B1	0.0	0.2	0.9	0.9
ingv_echam4	A2	0.6	0.8	1.0	0.0
	B1				
inmcm3_0	A2	0.4	0.6		
	B1	0.4	0.6		
ipsl_cm4	A2	0.0	0.6	0.4	0.9
	B1	0.1	0.6	0.5	0.9
miroc3_2_medres	A2	0.2	0.8		0.8
	B1	0.2	0.8		0.7
miub_echo_g	A2	0.9	1.0	1.0	0.9
	B1	0.9	1.0	1.0	0.9
mpi_echam5	A2	0.6	0.9	0.5	0.6
	B1	0.7	0.9	0.6	0.6
mri_cgcm2_3_2a	A2	1.0	1.0	0.4	0.8
	B1	1.0	1.0	0.4	0.7

Annual Mean Temperature

Inspection of Figure 2 – the B1 scenario – shows:

- a certain increase in mean annual temperature until 2050 (no overlap of the distribution with the present situation)
- a further increase of the expectation value until 2100, but a significant overlap with the distribution for 2050 – this further increase is probable but less certain than the increase until 2050

This is reflected in Table 4, B1-scenario, where between the expectation values for 2050 and 2100 exists an overlap from 28.6 to 28.8°C within the standard deviation ranges. Inspection Figure 2 – the A2 scenario – and the respective values in Table 4 show:

- until 2050 a certain increase compared to the present situation
- the expectation value in 2050 for the A2 scenario is slightly larger than for the B1-scenario, but this is uncertain due to the high overlap of the distributions
- there is a certain further increase until 2100 (very small overlap with the 2050 – distribution)
- the A2 increase until 2100 is certainly larger than the increase under the B1 – scenario

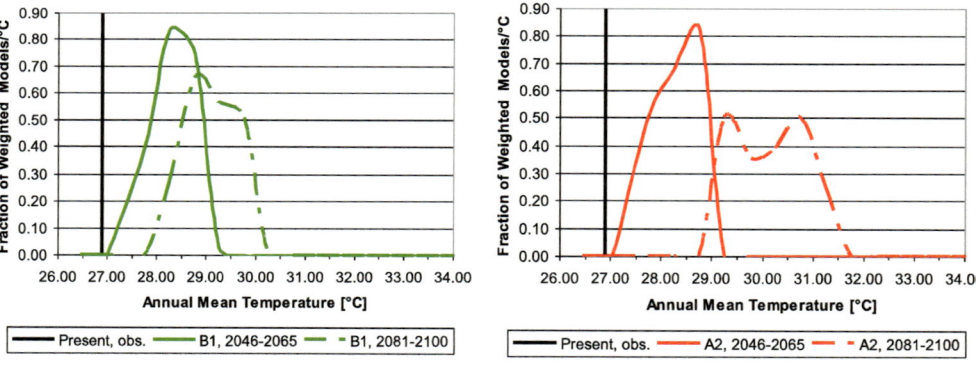

Figure 2: Coloured lines - Weighted Number of AOGCMs projecting the Mean Annual Temperature within a given interval. Black lines - Presently observed Annual Mean Temperature. l: B1 global GHG emission scenario; r: A2 global GHG emission scenario

Table 4: Weighted average of projected Mean Annual Temperature ± standard deviation [°C] for the different time slices and emission scenarios

Scenario:	Time slice:		
	Present	2046-2055	2081-2100
B1	26.9± 0.0	28.4± 0.4	29.1± 0.5
A2	26.9± 0.0	28.8± 0.4	30.7± 0.7

Annual Precipitation Sum

Inspection of Figure 3 and Table 5 shows a much less clear cut situation compared to the preceding climate variable. In line with the first results in our 2009-report the projections for the annual precipitation sum are highly diverse. We get:

- a slight, almost linear increase of the expectation value of the annual precipitation sum until 2100
- this increase of the expectation value is almost independent from the emission scenario
- but this increase is the average of a wide range of model results ranging from a significant decrease over time constancy to a significant increase
- the range of uncertainty is significantly larger for the high emission scenario

There are strong hints that the monsoon system, which is responsible for the total precipitation amount in this region, shows multi stability properties (Knopf et al., 2008; Zickfeld et al., 2005). It is likely that this highly non-linear behaviour is the reason for the diverting projections which would be in this case not an indicator of model uncertainty but intrinsic non-predictability of the monsoon system. In the light of this hypothesis the increase of the range of projected changes for the high emission case becomes an important result: B1 will keep the annual precipitation sum more predictable then A2.

Table 5: Weighted average of projected Annual Precipitation Sum ± standard deviation [mm] for the different time slices and emission scenarios

Scenario:	Time slice:		
	Present	2046-2055	2081-2100
B1	809± 0	852± 108	890± 133
A2	809± 0	853± 132	888± 207

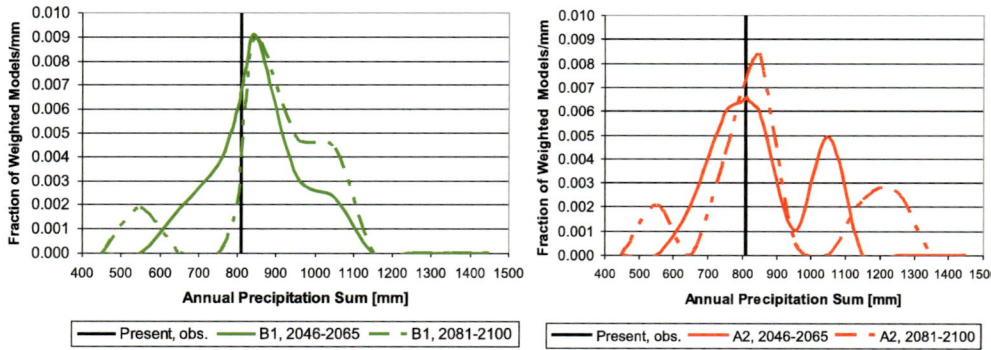

Figure 3: Coloured lines - Weighted Number of AOGCMs projecting the Annual Precipitation Sum within a given interval (100 mm-bins). Black lines - Presently observed Annual Precipitation Sum. l: B1 global GHG emission scenario; r: A2 global GHG emission scenario

Daily Precipitation

Inspection of Figure 4 (B1-emission scenario) and Table 6 shows:

- a relatively certain increase in greater 80mm/day precipitation days until 2050
- a considerable expectation value of 56 % increase
- a small probability exists that there will be no increase until 2050
- a small further increase to be expected until 2100 (reaching 76 %)
- for this time slice an increase compared to present is more certain than in 2050

Inspection of Figure 4 (A2-emission scenario) and Table 6 shows:

- an almost certain increase in greater 80 mm/day precipitation days until 2050 (non-increase outside the sigma-range)
- increase until 2050 comparable to B1-scenario (but more certain)
- a large further increase to be expected until 2100 (reaching 172 %)
- because of some overlap with the 2050 distribution there remains a small probability that there is no further increase until 2100

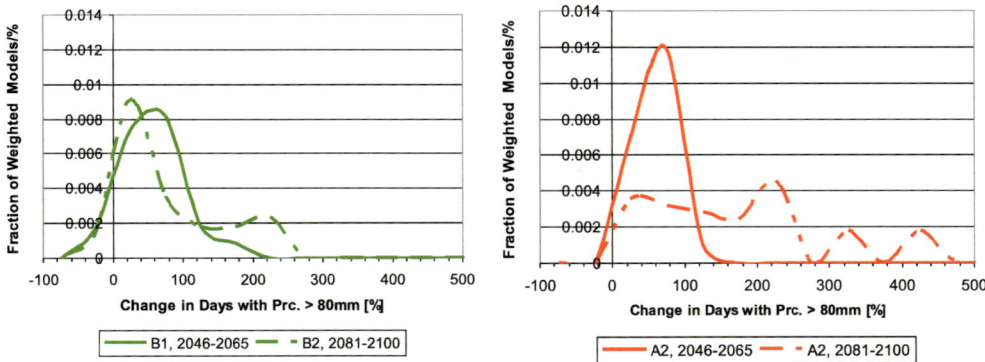

Figure 4: Coloured lines - Weighted Number of AOGCMs projecting the change of heavy rain days within a given interval (50 %-bins). l: B1 global GHG emission scenario; r: A2 global GHG emission scenario

Table 6: Weighted average of projected change in heavy rain days ± standard deviation [%] for the different time slices and emission scenarios

Scenario:	Time slice:	
	2046-2055	2081-2100
B1	57± 69	60± 39
A2	60± 39	172± 114

Heat Waves

Inspection of Figure 5 and Table 7 shows:

- a strong and certain increase in the expected number of heat wave days until 2050 under the B1 scenario

- a further, but weaker increase until 2100 in the expectation value of limited certainty (strong overlap of distributions)

- an extreme and certain increase in the expected number of heat wave days until 2050 under the A2 scenario to 18.9 days/year

- a further linear increase until 2100 to 41 days/year with somewhat lower certainty (strong overlap of distributions)

Table 7: Weighted average of projected Heatwave Days ± standard deviation [days] for the different time slices and emission scenarios

Scenario:	Time slice:		
	Present	2046-2055	2081-2100
B1	1.2± 0.0	8.0± 4.6	12.8± 7.4
A2	1.2± 0.0	18.9± 16.1	41.0± 21.4

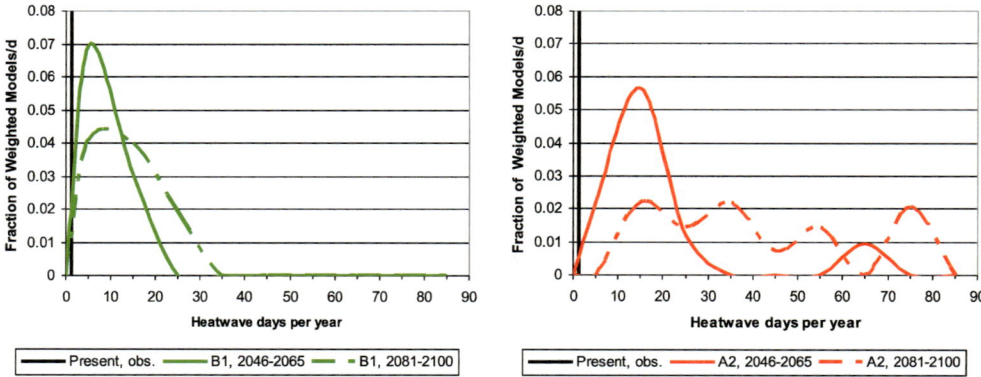

Figure 5: Coloured lines - Weighted Number of AOGCMs projecting the Heatwave Days within a given interval (10 days-bins). Black lines - Presently observed Heatwave Days. l: B1 global GHG emission scenario; r: A2 global GHG emission scenario

5 Discussion and Policy Implications

This paper delivers climate change projections for the urban area of Hyderabad (India) based on the full set of AOGCM results which were generated for the IPCC AR4 process. These projections include expectation values of climate change and estimations of their certainty taking convergence of model results as an indicator for their validity. This follows the underlying philosophy of IPCC AR4 which is in line with a "consensus theory of truth" and applies it – to our knowledge for the first time – to a regional climate change assessment.

We refer to runs for a high (A2) and a low (B1) global CO_2-emission scenario, the considered time slices are 1961-2000 (reference climate), 2046-2065 and 2081-2100 and evaluate these model runs to obtain projections of the four most impact-relevant climatic characteristics for Hyderabad: the annual mean temperature (e.g. for urban agriculture), the total annual precipitation (e.g. for urban water supply), the frequency distribution of daily precipitation (important, e.g, for urban flooding) and the probability of heat waves (e.g. for human health).

This work is a follow up of a former report which gave first estimations for the development of these climate variables. Main improvements are:

Algorithmic changes:

- change of the heat wave definition (adaptation to Indian situation)
- change of the downscaling algorithm for heavy rainfall (adaption to data quality)

Changes in the set evaluated AOGCM results:

- change to the full set of AOGCM results used in IPCC, AR4

The results obtained from the full analysis in this paper partly fostered our first estimations from the 09-report and partly demand for some modifications of the qualitative properties and quantitative values of the projections. As the model results were only a small subset of the results used in this report one would expect that the former expectation values and their uncertainty ranges lie within the uncertainty ranges calculated in this paper. But only in cases where the chosen subset was representative, the expectation value will be identical and only in cases were it spans the whole range of values the estimated uncertainty should be similar. This applies not necessarily for the heat wave projections as we switched here to a more appropriate definition.

For the projections of the mean annual temperature the qualitative properties given in the 09-report stay fully valid: it will develop monotonously in time and with a stronger trend in the high emission scenario up to +5°C. Also the conclusion still holds that this high value of +5°C for the A2 scenario in 2100 would definitely alter the natural water balance towards increased dryness, even under a (very uncertain) increase in total rainfall and a global emission reduction along the B1 scenario would clearly ease the adaptation pressure.

For the total annual precipitation in the region around Hyderabad in general our former best guess holds true that Hyderabad has to prepare for a change in average annual precipitation of about ±20 % and that it is neither possible to predict the sign nor the exact amount. We conclude now from secondary literature that this is not due to model uncertainties but to intrinsic unpredictability of the monsoon system. But with the full fledged analysis we can further detail this statement: the 20 % uncertainty range is valid except under the A2 scenario in 2100 – here a much larger uncertainty range of almost 40 % has to be dealt with. This is a further hint that the high emission case will lead to absolutely unmanageable situations.

For daily precipitation greater than 80 mm/day we estimated for the high emission scenario in the former report an increase in frequency of about 70 % (± 6) until 2050. This has to be corrected slightly to an expectation value of 60 %, although the uncertainty range is now larger (± 49 %). For the following time slice we have to correct our 2009 statement of a stabilization: this was an unrepresentative model behaviour and we know now that we have to expect an further increase to 172 %. Also in the low emission scenario we have to correct the qualitative course of the projection: until 2050 we will have an increase similar to the high emission case, but until 2100 there will only be a small further increase. So the preliminary conclusion drawn in the 2009-report that an immediate and effective reduction of global emissions will "buy some time" for adaptation is no longer valid. Instead we have to infer from the full fledged analysis that until 2050 we have to expect an increase in heavy rain events above 80 mm/day of about 50 % - mainly determined by historical emissions - while emission scenarios will become decisive in the second part of the century, again resulting in an alarming situation in the high emission case.

For the number of heat wave days a quantitative comparison with our former results is not reasonable (change of definition) but also the qualitative course of the projection is somewhat different: we conclude now that we have to expect for both emission scenarios a continuous increase in heat wave days during the century and that the low emission scenario results in significantly lower numbers of heat wave days compared to the high emission scenario. Again, the A2 scenario is expected to generate an alarming number of more than 40 heat wave days per year.

To summarise, this paper showed the different behaviour of the relevant climate variables in the Hyderabad area in response to global emission scenarios by using systematically thefull range of available global climate models. It allows thereby identifying adaptation necessities which are more or less dependent on the future development of the global greenhouse gas emissions.

6 References

Literature

Frich, A.; L.V. Alexander, P. Della-Marta, B. Gleason, M. Haylock, A.M.G. Klein Tank, and T. Peterson (January 2002). "Observed coherent changes in climatic extremes during the second half of the twentieth century" (PDF). Climate Research 19: 193–212. doi: 10.3354/cr019193. http://cccma.seos.uvic.ca/ETCCDMI/docs/Frichetal02.pdf

Hollweg, H.-D.; Böhm, U.; Fast, I.; Hennemuth, B.; Keuler, K.; Keup-Thiel, E.; Lautenschlager, M.; Legutke, S.; Radtke, K.; Rockel, B.; Schubert, M.; Will, A.; Woldt, M. and Wunram, C. (2008). Ensemble Simulations over Europe with the Regional Climate Model CLM forced with IPCC AR4 Global Scenarios. *Technical Report SGA-ZMAW*

Knopf, B., K. Zickfeld, M. Flechsig, and V. Petoukhov, 2008, Sensitivity of the Indian monsoon to human activities. *Advances of Atmospheric Sciences* 25(6): 932–945.

Meehl, G.A.; Stocker, T.F.; Collins, W.D.; Friedlingstein, P.; Gaye, A.T.; Gregory, J.M.; Kitoh, A.; Knutti, R.; Murphy, J.M.; Noda, A.; Raper, S.C.B.; Watterson, I.G.; Weaver, A.J. and Zhao, Z.-C. (2007). Global Climate Projections. In: Solomon, S.; Qin, D.; Manning, M.; Chen, Z.; Marquis, M.; Averyt, K.B.; Tignor, M. and Miller, H.L. (eds.). *Climate Change 2007: The Physical Science Basis. Contribution of Working Group I to the Fourth Assessment Report of the Intergovernmental Panel on Climate Change.* Cambridge University Press, Cambridge, United Kingdom and New York, NY, USA.

Mitchell, T.D.; Carter, T.R.; Jones, P.D.; Hulme, M.; New,M. (2007). A comprehensive set of high-resolution grids of monthly climate for Europe and the globe: the observed record (1901-2000) and 16 scenarios (2001-2100). Tyndall Centre Working Paper No. 55, July 2004.

Nakićenović, N. and Swart, R. (eds.) (2000): *Special Report on Emissions Scenarios: A Special Report of Working Group III of the Intergovernmental Panel on Climate Change.* Cambridge University Press, Cambridge, United Kingdom and New York, NY, USA, 599 pp.

Orlowsky, B., Gerstengarbe, F.-W. and Werner, P.C. (2008): A resampling scheme for regional climate simulations and its performance compared to a dynamical RCM. *Theor. Appl. Climatol.* 92(3-4): 209–223.

Shorak, G.R., Wellner, J.A. (1986) *Empirical Processes with Applications to Statistics*, Wiley.

Tebaldi, C.; Hayhoe, K.; Arblaster, J.M. and Meehl, G. (2006). Going to extremes. An intercomparison of model-simulated historical and future changes in extreme events. *Climatic Change* 79: 185–211.

Zickfeld, K., B. Knopf, V. Petoukhov, and H.-J. Schellnhuber, 2005, Is the Indian summer monsoon stable against global change?, *Geophysical Research Letters* 32, L15707.

Data Source

CERA / IPCC DDC (Climate and Environmental Retrieving and Archiving, DKRZ), http://cera-www.dkrz.de/CERA/index.html

Down to Earth Vol: 18 Issue: 20090531 pp: 1

IMD (India Meteorological Department)- National Data Centre, National Climate Centre, Shivajinagar, Pune – 411005, India

National Weather Service ; Silver Spring, Maryland, USA (25.05.2009). www.nws.noaa.gov/glossary/index.php?word=heat+advisory

xDat (eXtensible Database Access Tool). Potsdam Institute for Climate Impact Research. www.pik-potsdam.de/institute/organization/scientific-departments/data-computation/sdm/tools/xdat/

KNMI (Koninklijk Nederlands Meteorologisch Instituut)-Database, Wilhelminalaan 10, 3732 GK De Bilt, Netherlands

WCMIP3 World Climate Research Program's (WCRP's) Coupled Model Intercomparison Project phase 3 (CMIP3) multi-model dataset. We acknowledge the modeling groups for making their model output available for analysis, the Program for Climate Model Diagnosis and Intercomparison (PCMDI) for collecting and archiving this data, and the WCRP's Working Group on Coupled Modelling (WGCM) for organizing the model data analysis activity. The WCRP CMIP3 multi-model dataset is supported by the Office of Science, U.S. Department of Energy.

Appendix

Results of the single AOGCMs

Bjerknes Centre for Climate Research, Norway
BCCR-BCM2.0, 2005
bccr_bcm2_0

1. Mean annual temperature [°C]

Results for the respective grid element:

	1961-2000	2046-2065	2081-2100
B1	25.22	26.41	27.12
A2	25.22	26.75	28.90

Results scaled to Hyderabad:

	1961-2000	2046-2065	2081-2100
B1	26.88	28.15	28.92
A2	26.88	28.51	20.89

2. Annual precipitation sum [mm]

Results for the respective grid element:

	1961-2000	2046-2065	2081-2100
B1	1026.6	1115.1	1028.6
A2	1026.6	1115.6	1068.6

Results scaled to Hyderabad:

	1961-2000	2046-2065	2081-2100
B1	808.8	878.5	810.3
A2	808.8	878.9	841.8

3. Frequency of daily precipitation

Results for the respective grid element [days/year]

quantile [%]	0	80	90	94	97	99	99.86	99.98
1961-2000	291.26	36.87	14.75	11.07	7.36	3.16	0.44	0.08
A2, 2046-2065	288.29	35.21	16.62	11.18	10.43	2.72	0.45	0.10
A2, 2081-2100	296.57	32.56	14.10	9.69	7.08	3.37	1.33	0.29
B1, 2046-2065	290.31	37.14	14.63	9.95	7.84	3.92	1.01	0.20
B1, 2081-2100	296.51	34.17	11.67	10.87	7.78	2.82	0.96	0.21

4. Heat waves
Number of days/year fulfilling the heat wave day - definition
Results for the respective grid element:

	1961-2000	2046-2065	2081-2100
B1	0.375	6.85	61.25
A2	0.375	11.95	79.5

Canadian Centre for Climate Modelling and Analysis, Canada
CGCM3.1 (T47), 2005
cccma_cgcm3_1

1. Mean annual temperature [°C]

Results for the respective grid element:

	1961-2000	2046-2065	2081-2100
B1	24.65	26.45	27.05
A2	24.65	27.05	28.95

Results scaled to Hyderabad:

	1961-2000	2046-2065	2081-2100
B1	26.88	28.84	29.50
A2	26.88	29.50	31.57

2. Annual precipitation sum [mm]

Results for the respective grid element:

	1961-2000	2046-2065	2081-2100
B1	1002.85	1133.36	1222.62
A2	1002.85	1129.99	1443.55

Results scaled to Hyderabad:

	1961-2000	2046-2065	2081-2100
B1	808.8	914.1	986.0
A2	808.8	911.3	1164.2

3. Frequency of daily precipitation

Results for the respective grid element [days/year]

quantile [%]	0	80	90	94	97	99	99.86	99.98
1961-2000	292.00	36.50	14.60	10.95	7.30	3.15	0.43	0.08
A2, 2046-2065	286.90	36.85	17.00	11.85	7.65	3.85	0.75	0.15
A2, 2081-2100	280.60	39.30	16.45	12.55	8.75	5.20	1.35	0.80
B1, 2046-2065	290.30	36.10	14.25	12.15	7.05	4.20	0.85	0.10
B1, 2081-2100	290.10	33.55	15.25	11.75	8.35	4.70	1.00	0.30

4. Heat waves
Number of days/year fulfilling the heat wave day - definition
Results for the respective grid element:

	1961-2000	2046-2065	2081-2100
B1	1.175	4.8	7.45
A2	1.175	7.1	17.15

Canadian Centre for Climate Modelling and Analysis, Canada
CGCM3.1 (T63), 2005
cccma_cgcm3_1_t63

1. Mean annual temperature [°C]

Results for the respective grid element:

	1961-2000	2046-2065	2081-2100
B1	24.45	26.55	26.95
A2	X	X	X

Results scaled to Hyderabad:

	1961-2000	2046-2065	2081-2100
B1	26.88	29.19	29.63
A2	X	X	X

2. Annual precipitation sum [mm]

Results for the respective grid element:

	1961-2000	2046-2065	2081-2100
B1	1340.07	1591.70	1635.31
A2	X	X	X

Results scaled to Hyderabad:

	1961-2000	2046-2065	2081-2100
B1	808.8	960.7	987.0
A2	X	X	X

3. Frequency of daily precipitation

Results for the respective grid element [days/year]

quantile [%]	0	80	90	94	97	99	99.86	99.98
1961-2000	292.00	36.50	14.60	10.95	7.30	3.15	0.43	0.08
A2, 2046-2065								
A2, 2081-2100								
B1, 2046-2065	289.65	34.30	14.70	12.25	8.70	4.55	0.55	0.30
B1, 2081-2100	286.85	37.25	13.95	12.10	9.05	4.95	0.65	0.20

4. Heat waves
Number of days/year fulfilling the heat wave day - definition
Results for the respective grid element:

	1961-2000	2046-2065	2081-2100
B1	1.775	9.05	8.45
A2	X	X	X

Commonwealth Scientific and Industrial Research Organisation (CSIRO) Atmospheric Research, Australia
CSIRO-MK3.0, 2001
csiro_mk3_0

1. Mean annual temperature [°C]

Results for the respective grid element:

	1961-2000	2046-2065	2081-2100
B1	25.65	26.85	27.25
A2	25.65	27.35	29.15

Results scaled to Hyderabad:

	1961-2000	2046-2065	2081-2100
B1	26.88	28.14	28.56
A2	26.88	28.66	30.55

2. Annual precipitation sum [mm]

Results for the respective grid element:

	1961-2000	2046-2065	2081-2100
B1	603.92	651.69	672.74
A2	603.92	665.61	642.64

Results scaled to Hyderabad:

	1961-2000	2046-2065	2081-2100
B1	808.8	872.8	901.0
A2	808.8	891.4	860.6

3. Frequency of daily precipitation

Results for the respective grid element [days/year]

quantile [%]	0	80	90	94	97	99	99.86	99.98
1961-2000	292.00	36.50	14.60	10.95	7.30	3.15	0.43	0.08
A2, 2046-2065	291.30	35.10	12.90	12.75	8.05	4.10	0.50	0.30
A2, 2081-2100	291.70	34.75	14.20	12.15	7.80	3.55	0.50	0.35
B1, 2046-2065	291.25	34.05	13.90	13.30	8.50	3.25	0.65	0.10
B1, 2081-2100	289.10	36.50	15.00	12.00	7.40	4.30	0.40	0.30

4. Heat waves
Number of days/year fulfilling the heat wave day - definition
Results for the respective grid element:

	1961-2000	2046-2065	2081-2100
B1	3.85	7.15	11.6
A2	3.85	11.55	27.4

Commonwealth Scientific and Industrial Research Organisation (CSIRO) Atmospheric Research, Australia
CSIRO-MK3.5, 2010
csiro_mk3_5

1. Mean annual temperature [°C]

Results for the respective grid element:

	1961-2000	2046-2065	2081-2100
B1	29.55	31.45	32.05
A2	29.55	32.05	33.85

Results scaled to Hyderabad:

	1961-2000	2046-2065	2081-2100
B1	26.88	28.61	29.15
A2	26.88	29.15	30.79

2. Annual precipitation sum [mm]

Results for the respective grid element:

	1961-2000	2046-2065	2081-2100
B1	743.66	758.06	743.64
A2	743.66	705.67	799.24

Results scaled to Hyderabad:

	1961-2000	2046-2065	2081-2100
B1	808.8	824.5	808.8
A2	808.8	767.5	869.3

3. Frequency of daily precipitation

Results for the respective grid element [days/year]

quantile [%]	0	80	90	94	97	99	99.86	99.98
1961-2000	292.00	36.50	14.60	10.95	7.30	3.15	0.43	0.08
A2, 2046-2065	296.35	34.65	14.30	9.55	6.95	2.35	0.70	0.15
A2, 2081-2100	289.90	36.05	14.50	12.40	8.50	2.60	0.70	0.35
B1, 2046-2065	286.80	40.50	15.70	11.40	7.15	2.90	0.45	0.10
B1, 2081-2100	291.10	38.20	14.30	11.00	6.80	2.80	0.40	0.40

4. Heat waves
Number of days/year fulfilling the heat wave day - definition
Results for the respective grid element:

	1961-2000	2046-2065	2081-2100
B1	5.725	19.4	25
A2	5.725	25	50.25

U.S. Department of Commerce/National Oceanic and Atmospheric Administration (NOAA)/ Geophysical Fluid Dynamics Laboratory (GFDL), USA
GFDL-CM2.0, 2005
gfdl_cm2_0

1. Mean annual temperature [°C]

Results for the respective grid element:

	1961-2000	2046-2065	2081-2100
B1	25.15	26.75	27.35
A2	25.15	27.35	29.35

Results scaled to Hyderabad:

	1961-2000	2046-2065	2081-2100
B1	26.88	28.59	29.23
A2	26.88	29.23	31.37

2. Annual precipitation sum [mm]

Results for the respective grid element:

	1961-2000	2046-2065	2081-2100
B1	934.79	889.85	994.86
A2	934.79	865.53	842.40

Results scaled to Hyderabad:

	1961-2000	2046-2065	2081-2100
B1	808.8	769.9	860.8
A2	808.8	748.9	728.9

3. Frequency of daily precipitation

Results for the respective grid element [days/year]

quantile [%]	0	80	90	94	97	99	99.86	99.98
1961-2000	292.00	36.50	14.60	10.95	7.30	3.15	0.43	0.08
A2, 2046-2065	300.00	31.95	13.65	9.65	5.70	3.50	0.45	0.10
A2, 2081-2100	302.15	31.75	12.55	10.60	4.85	2.55	0.50	0.05
B1, 2046-2065	297.90	33.55	13.45	10.50	6.40	2.80	0.10	0.30
B1, 2081-2100	289.75	36.70	16.45	11.10	7.40	3.00	0.45	0.15

4. Heat waves
Number of days/year fulfilling the heat wave day - definition
Results for the respective grid element:

	1961-2000	2046-2065	2081-2100
B1	0.9	7.9	10.6
A2	0.9	10.75	34.3

U.S. Department of Commerce/National Oceanic and Atmospheric Administration (NOAA)/Geophysical Fluid Dynamics Laboratory (GFDL), USA
GFDL-CM2.1, 2005
gfdl_cm2_1

1. Mean annual temperature [°C]

Results for the respective grid element:

	1961-2000	2046-2065	2081-2100
B1	24.15	25.75	26.25
A2	24.15	26.25	28.35

Results scaled to Hyderabad:

	1961-2000	2046-2065	2081-2100
B1	26.88	28.66	29.22
A2	26.88	29.22	31.55

2. Annual precipitation sum [mm]

Results for the respective grid element:

	1961-2000	2046-2065	2081-2100
B1	1421.37	1598.07	1630.65
A2	1421.37	1487.84	1358.57

Results scaled to Hyderabad:

	1961-2000	2046-2065	2081-2100
B1	808.8	909.3	927.9
A2	808.8	846.6	773.1

3. Frequency of daily precipitation

Results for the respective grid element [days/year]

quantile [%]	0	80	90	94	97	99	99.86	99.98
1961-2000	292.00	36.50	14.60	10.95	7.30	3.15	0.43	0.08
A2, 2046-2065	298.45	30.10	13.50	10.95	6.80	3.95	0.80	0.45
A2, 2081-2100	310.85	24.20	10.35	7.60	6.50	3.90	1.10	0.50
B1, 2046-2065	293.55	32.05	15.15	11.10	7.25	4.55	1.15	0.20
B1, 2081-2100	297.65	29.85	13.55	9.20	7.60	5.40	1.35	0.40

4. Heat waves
Number of days/year fulfilling the heat wave day - definition
Results for the respective grid element:

	1961-2000	2046-2065	2081-2100
B1	3.25	15.45	19.25
A2	3.25	17.6	46.3

National Aeronautics and Space Administration (NASA)/Goddard Institute for Space Studies (GISS), USA
GISS-AOM, 2004
giss_aom

1. Mean annual temperature [°C]

Results for the respective grid element:

	1961-2000	2046-2065	2081-2100
B1	27.55	28.95	29.05
A2	X	X	X

Results scaled to Hyderabad:

	1961-2000	2046-2065	2081-2100
B1	26.88	28.25	28.34
A2	X	X	X

2. Annual precipitation sum [mm]

Results for the respective grid element:

	1961-2000	2046-2065	2081-2100
B1	763.23	757.18	899.33
A2	X	X	X

Results scaled to Hyderabad:

	1961-2000	2046-2065	2081-2100
B1	808.8	802.4	953.0
A2	X	X	X

3. Frequency of daily precipitation

Results for the respective grid element [days/year]

quantile [%]	0	80	90	94	97	99	99.86	99.98
1961-2000	292.00	36.50	14.61	10.94	7.30	3.15	0.42	0.07
A2, 2046-2065								
A2, 2081-2100								
B1, 2046-2065	293.50	29.28	16.49	14.69	7.54	2.65	0.85	0.00
B1, 2081-2100	281.20	33.08	17.14	16.29	11.99	4.60	0.60	0.10

4. Heat waves
Number of days/year fulfilling the heat wave day - definition
Results for the respective grid element:

	1961-2000	2046-2065	2081-2100
B1	0.475	6.65	8.5
A2	X	X	X

National Aeronautics and Space Administration (NASA)/Goddard Institute for Space Studies (GISS), USA
GISS-ER, 2004
giss_model_e_r

1. Mean annual temperature [°C]

Results for the respective grid element:

	1961-2000	2046-2065	2081-2100
B1	28.05	29.95	30.55
A2	28.05	30.45	32.35

Results scaled to Hyderabad:

	1961-2000	2046-2065	2081-2100
B1	26.88	28.70	29.28
A2	26.88	29.18	31.00

2. Annual precipitation sum [mm]

Results for the respective grid element:

	1961-2000	2046-2065	2081-2100
B1	730.64	599.39	535.35
A2	730.64	571.99	495.95

Results scaled to Hyderabad:

	1961-2000	2046-2065	2081-2100
B1	808.8	663.5	592.6
A2	808.8	633.2	549.0

3. Frequency of daily precipitation

Results for the respective grid element [days/year]

quantile [%]	0	80	90	94	97	99	99.86	99.98
1961-2000	292.00	36.50	14.60	11.00	7.30	3.10	0.40	0.10
A2, 2046-2065	298.40	40.80	10.95	8.05	4.20	2.50	0.00	0.10
A2, 2081-2100	304.75	38.80	9.85	5.90	3.70	1.90	0.05	0.05
B1, 2046-2065	296.95	40.35	13.00	6.90	4.65	3.00	0.05	0.10
B1, 2081-2100	298.70	42.00	11.70	6.65	3.65	1.95	0.35	0.00

4. Heat waves
Number of days/year fulfilling the heat wave day - definition
Results for the respective grid element:

	1961-2000	2046-2065	2081-2100
B1	3.475	7.1	21.3
A2	3.475	19.1	52.35

National Key Laboratory of Numerical Modelling for Atmospheric Sciences and Geophysical Fluid Dynamics (LASG)/Institute of Atmospheric Physics, China
FGOALS-g1.0, 2004
iap_fgoals1_0_g

1. Mean annual temperature [°C]

Results for the respective grid element:

	1961-2000	2046-2065	2081-2100
B1	27.65	28.75	29.35
A2	x	x	x

Results scaled to Hyderabad:

	1961-2000	2046-2065	2081-2100
B1	26.88	27.95	28.53
A2	x	x	x

2. Annual precipitation sum [mm]

Results for the respective grid element:

	1961-2000	2046-2065	2081-2100
B1	1538.78	1566.72	1494.19
A2	x	x	x

Results scaled to Hyderabad:

	1961-2000	2046-2065	2081-2100
B1	808.8	823.5	785.4
A2	x	x	x

3. Frequency of daily precipitation

Results for the respective grid element [days/year]

quantile [%]	0	80	90	94	97	99	99.86	99.98
1961-2000	292.00	36.50	14.60	10.95	7.30	3.15	0.43	0.08
A2, 2046-2065								
A2, 2081-2100								
B1, 2046-2065	290.30	33.70	15.75	12.25	8.75	3.95	0.30	0.00
B1, 2081-2100	297.25	31.95	13.05	10.15	7.45	4.75	0.20	0.20

4. Heat waves
Number of days/year fulfilling the heat wave day - definition
Results for the respective grid element:

	1961-2000	2046-2065	2081-2100
B1	0.925	2.85	6.15
A2	x	x	x

National Institute of Geophysics and Vulcanology (INGV), Italy
INGV-ECHAM4, 2006
ingv_echam4

1. Mean annual temperature [°C]

Results for the respective grid element:

	1961-2000	2046-2065	2081-2100
B1	X	X	X
A2	27.15	29.15	30.55

Results scaled to Hyderabad:

	1961-2000	2046-2065	2081-2100
B1	X	X	X
A2	26.88	28.86	30.25

2. Annual precipitation sum [mm]

Results for the respective grid element:

	1961-2000	2046-2065	2081-2100
B1	X	X	X
A2	1144.86	1109.25	1130.69

Results scaled to Hyderabad:

	1961-2000	2046-2065	2081-2100
B1	X	X	X
A2	808.8	783.6	798.8

3. Frequency of daily precipitation

Results for the respective grid element [days/year]

quantile [%]	0	80	90	94	97	99	99.86	99.98
1961-2000	292.00	36.50	14.60	10.95	7.30	3.14	0.43	0.08
A2, 2046-2065	297.07	32.44	14.04	10.34	6.74	3.70	0.66	0.00
A2, 2081-2100	296.41	30.72	13.74	11.15	8.31	4.11	0.56	0.00
B1, 2046-2065								
B1, 2081-2100								

4. Heat waves
Number of days/year fulfilling the heat wave day - definition
Results for the respective grid element:

	1961-2000	2046-2065	2081-2100
B1	X	X	X
A2	6.025	26.55	43.1

Institute for Numerical Mathematics, Russia
INM-CM3.0, 2004
inmcm3_0

1. Mean annual temperature [°C]

Results for the respective grid element:

	1961-2000	2046-2065	2081-2100
B1	x	x	x
A2	x	x	x

Results scaled to Hyderabad:

	1961-2000	2046-2065	2081-2100
B1	x	x	x
A2	x	x	x

2. Annual precipitation sum [mm]

Results for the respective grid element:

	1961-2000	2046-2065	2081-2100
B1	1279.63	1383.34	1339.21
A2	1279.63	1506.88	1382.53

Results scaled to Hyderabad:

	1961-2000	2046-2065	2081-2100
B1	808.8	874.3	846.5
A2	808.8	952.4	873.8

3. Frequency of daily precipitation

Results for the respective grid element [days/year]

quantile [%]	0	80	90	94	97	99	99.86	99.98
1961-2000	291.95	36.53	14.60	10.95	7.32	3.15	0.43	0.08
A2, 2046-2065	279.45	33.95	16.70	15.40	13.00	5.70	0.70	0.10
A2, 2081-2100	290.70	27.55	12.55	14.10	11.70	7.10	1.10	0.20
B1, 2046-2065	288.80	33.45	14.85	13.40	8.10	5.55	0.80	0.05
B1, 2081-2100	292.05	30.30	13.05	11.95	11.75	5.25	0.65	0.00

4. Heat waves
Number of days/year fulfilling the heat wave day - definition
Results for the respective grid element:

	1961-2000	2046-2065	2081-2100
B1	x	x	x
A2	x	x	x

Institut Pierre Simon Laplace, France
IPSL-CM4, 2005
ipsl_cm4

1. Mean annual temperature [°C]

Results for the respective grid element:

	1961-2000	2046-2065	2081-2100
B1	28.85	30.85	31.45
A2	28.85	31.25	33.55

Results scaled to Hyderabad:

	1961-2000	2046-2065	2081-2100
B1	26.88	28.74	29.30
A2	26.88	26.88	29.12

2. Annual precipitation sum [mm]

Results for the respective grid element:

	1961-2000	2046-2065	2081-2100
B1	252.78	257.04	265.22
A2	252.78	197.22	289.29

Results scaled to Hyderabad:

	1961-2000	2046-2065	2081-2100
B1	808.8	822.4	848.6
A2	808.8	631.0	925.6

3. Frequency of daily precipitation

Results for the respective grid element [days/year]

quantile [%]	0	80	90	94	97	99	99.86	99.98
1961-2000	292.00	36.50	14.60	10.95	7.30	3.14	0.43	0.08
A2, 2046-2065	308.22	31.23	10.70	6.49	5.58	1.93	0.66	0.20
A2, 2081-2100	293.32	38.17	12.17	9.78	6.89	3.55	0.61	0.51
B1, 2046-2065	299.86	33.00	12.37	8.36	6.84	3.60	0.66	0.30
B1, 2081-2100	293.83	37.01	11.61	9.38	7.00	5.63	0.46	0.10

4. Heat waves
Number of days/year fulfilling the heat wave day - definition
Results for the respective grid element:

	1961-2000	2046-2065	2081-2100
B1	0.575	14	27.1
A2	0.575	21.45	77.1

Center for Climate System Research (University of Tokyo), National Institute for Environmental Studies, and Frontier Research Center for Global Change (JAM-STEC), Japan
MIROC3.2(medres), 2004
miroc3_2_medres

1. Mean annual temperature [°C]

Results for the respective grid element:

	1961-2000	2046-2065	2081-2100
B1	x	x	x
A2	x	x	x

Results scaled to Hyderabad:

	1961-2000	2046-2065	2081-2100
B1	x	x	x
A2	x	x	x

2. Annual precipitation sum [mm]

Results for the respective grid element:

	1961-2000	2046-2065	2081-2100
B1	1389.78	1443.99	1480.48
A2	1389.78	1430.24	1566.77

Results scaled to Hyderabad:

	1961-2000	2046-2065	2081-2100
B1	808.8	840.4	861.6
A2	808.8	832.4	911.8

3. Frequency of daily precipitation

Results for the respective grid element [days/year]

quantile [%]	0	80	90	94	97	99	99.86	99.98
1961-2000	292.00	36.50	14.61	10.94	7.30	3.15	0.42	0.07
A2, 2046-2065	293.25	36.03	12.99	9.74	7.89	4.30	0.80	0.00
A2, 2081-2100	287.49	34.13	16.19	12.14	9.69	4.35	0.80	0.20
B1, 2046-2065	291.40	33.93	14.89	11.89	9.09	3.25	0.55	0.00
B1, 2081-2100	288.49	35.18	16.24	12.54	8.85	3.25	0.40	0.05

4. Heat waves
Number of days/year fulfilling the heat wave day - definition
Results for the respective grid element:

	1961-2000	2046-2065	2081-2100
B1	0.025	13	17.55
A2	0.025	10.3	27.25

Meteorological Institute of the University of Bonn, Meteorological Research Institute of the Korean Meteorological Agency, and Model and Data group, Germany/Korea
MIUB/ECHO-G, 2005
miub_echo_g

1. Mean annual temperature [°C]

Results for the respective grid element:

	1961-2000	2046-2065	2081-2100
B1	27.05	28.25	29.05
A2	27.05	28.45	30.15

Results scaled to Hyderabad:

	1961-2000	2046-2065	2081-2100
B1	26.88	28.07	28.87
A2	26.88	28.27	29.96

2. Annual precipitation sum [mm]

Results for the respective grid element:

	1961-2000	2046-2065	2081-2100
B1	782.27	944.63	1480.48
A2	782.27	970.42	1566.77

Results scaled to Hyderabad:

	1961-2000	2046-2065	2081-2100
B1	808.8	976.7	1003.3
A2	808.8	1002.5	1131.6

3. Frequency of daily precipitation

Results for the respective grid element [days/year]

quantile [%]	0	80	90	94	97	99	99.86	99.98
1961-2000	292.00	36.50	14.60	10.95	7.30	3.14	0.43	0.08
A2, 2046-2065	275.98	43.14	18.71	12.22	10.19	4.06	0.61	0.10
A2, 2081-2100	270.66	43.45	17.24	16.32	10.19	5.48	1.37	0.30
B1, 2046-2065	279.73	41.27	15.77	14.04	9.23	4.11	0.76	0.10
B1, 2081-2100	279.33	41.67	15.11	13.38	9.07	4.77	1.32	0.35

4. Heat waves
Number of days/year fulfilling the heat wave day - definition
Results for the respective grid element:

	1961-2000	2046-2065	2081-2100
B1	0.725	6.1	15.05
A2	0.725	11.8	30.45

Max-Planck-Institute for Meteorology, Germany
ECHAM5-OM, 2005
mpi_echam_5

1. Mean annual temperature [°C]

Results for the respective grid element:

	1961-2000	2046-2065	2081-2100
B1	25.75	27.75	28.65
A2	25.75	27.85	30.55

Results scaled to Hyderabad:

	1961-2000	2046-2065	2081-2100
B1	26.88	28.97	29.91
A2	26.88	29.07	31.89

2. Annual precipitation sum [mm]

Results for the respective grid element:

	1961-2000	2046-2065	2081-2100
B1	1110.87	1034.97	1118.61
A2	1110.87	1118.52	1168.21

Results scaled to Hyderabad:

	1961-2000	2046-2065	2081-2100
B1	808.8	753.5	814.4
A2	808.8	814.4	850.6

3. Frequency of daily precipitation

Results for the respective grid element [days/year]

quantile [%]	0	80	90	94	97	99	99.86	99.98
1961-2000	292.00	36.50	14.61	10.94	7.30	3.15	0.42	0.07
A2, 2046-2065	297.80	31.18	13.84	11.14	6.70	3.45	0.65	0.25
A2, 2081-2100	298.64	30.28	13.44	10.14	7.35	3.80	0.75	0.60
B1, 2046-2065	297.55	33.93	13.19	10.34	7.00	2.35	0.50	0.15
B1, 2081-2100	295.09	32.83	14.24	11.59	7.00	3.65	0.45	0.15

4. Heat waves
Number of days/year fulfilling the heat wave day - definition
Results for the respective grid element:

	1961-2000	2046-2065	2081-2100
B1	3.15	15.75	28.8
A2	3.15	18.75	51.55

Meteorological Research Institute, Japan
MRI-CGCM2.3.2, 2003
mri_cgcm_2_3_2a

1. Mean annual temperature [°C]

Results for the respective grid element:

	1961-2000	2046-2065	2081-2100
B1	25.45	26.35	27.05
A2	25.45	26.55	28.05

Results scaled to Hyderabad:

	1961-2000	2046-2065	2081-2100
B1	26.88	27.83	28.57
A2	26.88	28.04	29.63

2. Annual precipitation sum [mm]

Results for the respective grid element:

	1961-2000	2046-2065	2081-2100
B1	896.67	1134.39	1146.37
A2	896.67	1146.70	1408.73

Results scaled to Hyderabad:

	1961-2000	2046-2065	2081-2100
B1	808.8	1023.2	1034.0
A2	808.8	1034.3	1270.7

3. Frequency of daily precipitation

Results for the respective grid element [days/year]

quantile [%]	0	80	90	94	97	99	99.86	99.98
1961-2000	292.00	36.50	14.60	10.95	7.30	3.15	0.43	0.08
A2, 2046-2065	277.15	38.75	18.45	14.00	11.10	4.60	0.75	0.20
A2, 2081-2100	265.60	40.75	20.30	14.90	13.35	7.50	2.10	0.50
B1, 2046-2065	277.15	38.80	18.30	14.35	10.35	5.30	0.75	0.00
B1, 2081-2100	274.50	42.10	19.50	12.85	10.45	4.85	0.70	0.05

4. Heat waves
Number of days/year fulfilling the heat wave day - definition
Results for the respective grid element:

	1961-2000	2046-2065	2081-2100
B1	0	0.8	2.95
A2	0	0.55	10.7